# A LOOMING FLOOD

## IN THE BLUE

JOHN MACKSON

Copyright © 2022 John Mackson

All Rights Reserved

# TABLE OF CONTENTS

| | |
|---|---|
| **INTRODUCTION** | 4 |
| **CHAPTER ONE** | 6 |
| **FLOODING** | 6 |
| **FLOODS TYPES** | 6 |
| **FLOODS AND THEIR CAUSES** | 7 |
| **FLOODS AND THEIR IMPACT** | 8 |
| FLOODING'S AGRICULTURAL CONSEQUENCES | 9 |
| FLOODING HAS A NEGATIVE IMPACT ON HOUSES. | 10 |
| **MANAGEMENT OF FLOODING** | 11 |
| **CHAPTER TWO** | 18 |
| YELLOWSTONE | 18 |
| **CHAPTER THREE** | 34 |
| YELLOWSTONE FLOODING. | 34 |
| **CHAPTER FOUR.** | 54 |
| **FLASH FLOOD** | 54 |
| CONCERNING FLASH FLOODS | 57 |
| **CHAPTER FIVE** | 62 |
| HOW QUICKLY MAY FLASH FLOODS HAPPEN? | 62 |
| WHAT CAUSED THE FLASH FLOOD TO OCCUR? | 63 |

| | |
|---|---|
| **CHAPTER SIX** | **76** |
| FLOOD WARNING. | 76 |
| **CONCLUSION** | **94** |
| YELLOWSTONE RIVER | 94 |

# INTRODUCTION

When rain falls at a rate of more than 50 mm per square meter, it is probable that flooding will occur if the rain pools. This is a dramatic rise in water level in a short period of time in a location where there was previously no water.

It can occur anywhere, but some areas are more vulnerable to the occurrence than others.

Under the intense wind and heavy rain, such as during a hurricane, the sea may flood the beaches, causing the sea to rise a few meters.

Due to excessive rain, a river or stream may see an upsurge of water and escape its bed.

In just one hour, we can get 400 mm of water per square meter. When this occurs, there are significant disasters and fatalities.

They are not common, but they do occur, and the threat is sometimes understood, yet people go about their lives until they are forced to confront it.

Flooding occurs when a considerable amount of water exceeds its normal bounds, particularly over ordinarily dry territory. One of the most harmful, expensive, and common natural disasters in the world is flooding.

# CHAPTER ONE

## FLOODING

Flooding occurs when the water level in a river channel or along the shore rises to a point where it inundates land that is not ordinarily inundated. Flood is a physical feature of the environment and thus a crucial part of a drainage basin's hydrological cycle.

Flooding is a natural occurrence that occurs in reaction to excessive rainfall, but it becomes a hazard when it causes damage to people's lives and property.

## FLOODS TYPES

Riverine, estuarine, Coastal, and Manmade floods are the four main types of flooding.

Riverine floods are classified into two types: slow-rise floods generated by rain or snowmelt, and more rapid flash floods caused primarily by severe thunderstorms.

Estuarine floods are typically generated by a combination of a tidal surge at sea caused by storm-force winds and river flooding caused by inland rainstorms.

Tropical cyclones and other extreme sea storms, as well as tsunami waves, produce coastal flooding. Finally, several catastrophic reasons, such as dam failure or the consequences of an earthquake or volcanic eruption, can be identified.

Man-made floods are created by anthropogenic activities that disrupt the natural hydrological cycle, resulting in the inundation of areas that are not normally flooded.

# FLOODS AND THEIR CAUSES

Several factors contribute to floods, including: Heavy rain, cloud bursts, and cyclones are all examples of meteorological conditions.

Narrow exits, large catchment areas, a lack of well-developed drainage channels, channel siltation and rising, and the presence of unconsolidated soil are all physical conditions. Landslides' blocking effect, Meandering

Human impact can be seen in the construction of dams and reservoirs, dam bursts, deforestation, incorrect slope operations, and embankment construction.

# FLOODS AND THEIR IMPACT

Floods have medical, public health, and dietary repercussions.

Floods have a greater impact on human health and nutrition in less developed countries.

Obviously, death is the most catastrophic health consequence. It usually happens as a result of drowning, impact (falling trees, collapsing buildings, etc.), or exposure and solitude. Physical trauma and accompanying wound infection are both common causes of injury.

Other health impacts include respiratory disorders caused by exposure, gastrointestinal diseases caused by poor sanitation and water supplies, diseases linked to ecological changes (in rodent and insect habitats, for example), and malnutrition with high newborn mortality rates (When food sources are destroyed).

# Flooding's Agricultural Consequences

Flooding, particularly that caused by cyclones, can cause significant damage to the agriculture sector of the economy, which may be the primary source of domestic and international revenue.

High winds, torrential rains, and storm surges associated with cyclones that make landfall in coastal agricultural areas can cause defoliation, stripping, and cracking of limbs, trunks, and stems, among other things. twisting and whipping back stems, as well as lifting and exposing roots, especially in shallow soils and root systems

Lower-growing crops may be damaged or suffocated by falling branches and trees, while flooding lasting more than one or two days may kill crops and hide their remains with deposited sediment.

Crop and livestock losses; seed, rootstock, or breeding stock losses; income and employment losses; and loss of production potential are the main effects of floods on agriculture.

## Flooding has a negative impact on houses.

Floods and cyclones can have a significant impact on housing.

1. While all buildings deteriorate over time, repeated floods can hasten the process by

causing salt effervescence, damp rot, and other detrimental effects.

2. Second, because its center of buoyancy is higher than its center of gravity, a wooden structure will float upright.

3. Concrete strip foundations are more easily displaced because the weight of concrete below water is reduced by half.

4. Finally, if air pockets trapped in a building's cavities are compressed, they might cause deadly bursts.

# MANAGEMENT OF FLOODING

## 1. Runoff reduction

Reduced runoff is achieved by inducing and increasing infiltration into the earth in the catchment area, which is the most effective method of flood management. This can be accomplished through large-scale afforestation with trees that produce more trash.

## 2. Using volume reduction to reduce flood peaks

Flood peaks can be lowered by reducing the volume of water, which can be accomplished by engineering methods such as the construction of storage reservoirs and detention basins. During floods, such storage reservoirs impound massive amounts of water, reducing the amount of water carried by rivers.

## 3. Reducing flood levels is number

Flood levels can be decreased in a number of ways.

(a) The channelization of a stream. As the flood waves proceed downstream, the canals function as temporary storage and store water. As a result, the flood's severity is decreased.

(b) Enhancement of the channel. The flood conveyance capacity of rivers is increased by deepening, broadening, straightening, lining, and clearing vegetation and debris from the channels.

c) Diversion of floodwaters Flood diversion entails diverting and distributing flood water over railway lines and roadways, as well as sparsely over paddy fields and desert drylands.

## 4. Protection from Inundation

To protect the area from flooding, embankments are built to prevent water from pouring into inhabited or farmed areas.

## 5. Zoning in flood plains (FPZ)

Floodways are identified in connection to land use. After researching flood cycles, precise maps of flood-prone locations are created. Some locations are more prone to flooding, whereas floods occur more frequently in others. These zones are recognized and delineated, and essential land-use control is exercised.

## 6. Flood Prediction

Flood information that is available ahead of time is also important. Flood losses, particularly human life and animals, can be significantly reduced if flood predictions and early warnings are provided to the impacted communities.

## 7. Diversion of Water

The government has also considered redirecting excess water from flooding rivers to areas that are prone to drought. The Himalayan rivers would be linked to those in Peninsular India, according to the national water resource development strategy.

To preserve lives, "flash" floods necessitate a fast response. (Both flash flooding and "non-flash floods" cause property damage.)

The wording "flash" adds a crucial element of time to floods. Sudden, catastrophic dam failures or torrential rainfall in a tiny, steep drainage basin where the earth can't absorb the rain are obvious examples of "flash" flooding.

Flash flooding, on the other hand, does not always imply a "wall of water" rushing down a steep slope. It also involves swiftly rising water caused by a local drainage basin being overwhelmed by unusually heavy rain. When a slow-moving heavy rain producer, such as a tropical storm, strikes a low-lying metropolitan environment, this happens more frequently.

When the leftovers of Tropical Storm Allison struck Tallahassee in 2001, this happened. (When the storm stalled over Houston, it had already caused extensive, catastrophic flash flooding and "non-flash" flooding.) That night, I drove to work at the National Weather Service office during some of the strongest rainstorms I'd ever seen. Our weather station in Tallahassee received almost 8 inches of rain during my 8-hour shift, with the majority of it falling during the first half of my night shift.

Rainfall rates of 3–6 inches per hour can soon overwhelm even the best urban drainage systems. The water then rises at such a rapid rate that residents in these locations are forced to flee to higher land. Vehicles frequently sink or float away. Lives are in peril if a timely and appropriate response is not provided.

in traditionally flood-prone metropolitan regions may be unconcerned about the threat because they are used to it. What they aren't used to is how quickly this flooding happens. This type of flash flooding can elevate water levels to record levels in just an hour or two.

The key time factor is missing in a "non-flash" flood. Floods are primarily a threat to cars attempting to cross through flooded areas. When the upstream river crest wave slowly makes its way downstream, flood warnings are often issued along river flood plains. Residents in these areas normally have plenty of time to prepare for property damage, and there should be no need for human life to be lost in such circumstances.

To summarize, both types of flooding are destructive, but "non-flash" flooding provides far more time for flood damage and loss of life to be mitigated. Flash flooding should be regarded as any other weather emergency; if you are in the affected area, you must act quickly and correctly to ensure your safety.

# CHAPTER TWO

## YELLOWSTONE

Yellowstone is the epitome of a movement that altered the course of history!

Parks served a very different purpose in the globe before a part of what is now Yosemite was set aside as a state of California preserve in 1864. They were wealthy and/or nobility's hunting grounds or places to enjoy the fruits of their labor.

For the first time in 1872, a place was set aside for the enjoyment of everybody.

has influenced and inspired people all around the world, resulting in thousands of national parks where nature is protected for its inherent value and everyone can visit to recharge their spiritual essence! What happened first in Yellowstone has a name: egalitarian. This place is yours and mine, as well as the creatures and plants that live there.

And what a location! It is the lower 48 United States' biggest intact ecosystem, with all of the essential players interacting with one another and the land as they have for millennia! It is appropriately referred to as a gigantic museum of how things were 300 or more years ago across the continent.

Because the earth's crust above volcanism is as thin as it gets, the ancient forces that produced the globe are clearly apparent here! The hidden beauty and force of hydrothermal explosions astound in higher numbers than anywhere else on Earth!

Mineral deposits and microbial life forms, on the other hand, have a delicate delicacy that harkens back to the dawn of life on this planet. Amazing waterfalls in magnificent canyons inspire viewers' imaginations to soar with the osprey that nest on spires within them. Exotic wildlife thrives because there is so much land away from the road that they may not be seen until they are, in serendipitous majesty, to humble and amaze us.

The Park has been designed to allow curious visitors access, but the vast majority of the park remains undeveloped in order to preserve its natural beauty. It's a land so isolated and unknown that sojourners can get lost in its heart and emerge with newfound vigor and purpose.

Yellowstone is a breath-taking jewel woven into a fabric of wonder that spans the globe

Wyoming is home to 96 percent of Yellowstone National Park. The scientists who chronicled its wonders advised against settlement and development based just on the location, regardless of whatever region it was in.

Many people believe there is more than 3% in Montana because of strong marketing, but there isn't. Except for the park's bulging northwest corner, a four-kilometer-wide strip runs across the park's northern and western edges until it meets the Idaho state line. In that state, the four-kilometer strip extends south, containing the remaining one percent.

Exploring Yellowstone, or any other national park or forest, rather than merely visiting, is a transformative experience. "But it's already been investigated," someone may say. Not yet, not by you or me. And, since it was initially cataloged a century and a half ago, much of it has been rarely seen.

By the way, seeing the highlights is a relative exploration if it's your first, second, or third visit to Yellowstone. I didn't mean to come out as arrogant; I adore everything. However, I interpreted "explore" to mean "go where the throngs don't go."

The major attractions are enjoyable, but only on the surface. It's magical to sit beside a little-known creek or discover a secret lake or small waterfall. Serenity can be found in the smell of the air, in unexpected sounds, or in the presence of wildlife; it is Yellowstone's hidden heart.

Large swaths of the park and the wilderness areas surrounding it are devoid of trails. It's possible that First Peoples traveled all the way around Yellowstone Lake long ago, but when the 1870 Washburn expedition circumnavigated it, there was no evidence of previous transit.

Members of that group, as well as the subsequent Hayden surveys, climbed virgin summits and gazed down on vast swaths of land that, at the time, and now, have little reason to be visited.

Bushwhacking is as bit as exhausting as it sounds, and not in the sense of waylaying someone. However, it opens up the great majority of our public lands, particularly in our larger parks and forests, where the majority of the land is set aside for animal habitat, vegetation, and geology and is almost inaccessible. You are invited to visit, but the admittance ticket is toil and sweat, as it was two centuries ago.

You can get farther away from even a dirt road in the area marked T for Thoroughfare above than anywhere else in the contiguous United States.

Please make sure you know what you're doing if you venture into the backcountry. Make sure you have the proper permits and equipment, as well as the knowledge to use it.

90% of Yellowstone is not visible from the road. Even if you walked all 1,700 kilometers of hiking paths in the park, you would still miss out on the majority of it. This is a treasure trove and a museum of how the entire continent used to be.

I'm sure some interested First Americans, whether visionary or shaman, traveled to impractical areas just to see and feel them, but if that was the case, you'll discover no proof of their presence when you were there. In previous blogs, I argued that national parks and wilderness regions were established for non-utilitarian purposes; they exist to nourish your souls rather than your physical requirements. All of the challenges and rigors of previous eras, as well as the awe and solitude, are still there to be experienced.

You can see a long way from high elevations. You can see, metaphorically, how you got here and where we might go. You can also see an endless horizon beyond that, which is full of further unknowns. Perhaps, like ancient holy people, you can convey your vision to the world of humanity so that everyone benefits. Other areas are densely covered in dense vegetation, giving the impression that you can't see anything.

Look around again; you're in the midst of primal splendor. Sunbeams between the trees cast a glow similar to that seen between the columns of European churches.

Your unique insights are revealed through personal exploration. I have little doubt that every acre of Yellowstone has been visited, despite the fact that much of it has changed since the initial visit. Most of it is a relative frontier for you and me, beyond the horizon or past the ridgetop beyond the bucket list attractions.

It is a new world for us to explore on our own. And the more we investigate and go to see, the more we understand our own place in the cosmos and our own domain, allowing us to feel at peace within ourselves.

Yellowstone National Park is closed due to record flooding and mudslides.

Visitors will be unable to enter the park, which spans portions of Wyoming, Montana, and Idaho, while officials assess damage to roads and bridges.

The Park superintendent announced that due to record floods and rockslides caused by a burst of torrential rainfall, all five entrances to Yellowstone national park were closed on Monday, the first day of the summer tourist season.

The entire park, which includes parts of Wyoming, Montana, and Idaho, will be closed to visitors, including those with accommodation and camping reservations, at least until Wednesday as officials analyze the situation.

For the first time since a series of deadly wildfires in 1988, all five park entrances were blocked to inbound vehicles. Officials said the National Park Service was seeking to contact visitors and personnel who had remained in various locations, particularly in Yellowstone's hardest-hit northern side.

In a statement, park superintendent Cam Sholly stated, "Our main priority has been to evacuate the northern section of the park, where we have several road and bridge failures, mudslides, and other concerns."

The Gardiner River, which flows past Yellowstone National Park's north entrance, is high enough to have flooded part of the road. Photo by the Associated Press

A mudslide has blocked off the "gateway" community of Gardiner, Montana, which is located just north of the park's northern boundary and is home to many Yellowstone service workers, according to the National Park Service. The park's electric power was knocked out in multiple parts, according to the park service, and preliminary inspections revealed substantial stretches of road were washed away or covered in boulders and mud, as well as a number of bridges.

According to the group, several roads in the park's southern tier were on the verge of being swamped.

Days of intense rain in the park, as well as steady precipitation across most of the region, triggered the flooding and avalanche after one of the wettest springs in recent years. The Park authorities described the rain and flooding that swept over the area as "extraordinary."

An unexpected surge in summer temperatures over the last three days has expedited the melting and flow of late-winter storm-related snow at the park's highest altitudes.

Just two weeks after the start of the US summer tourist season on Memorial Day weekend, which accounts for the majority of Yellowstone's 4 million annual visitors, heavy rains and rapid snowmelt flow coincided in the park.

Flooding and mudslides prompted Yellowstone to close for the first time in 34 years.

Reuters, CODY, Wyo., June 13 - According to the park superintendent, all five entrances to

Yellowstone National Park were closed on Monday, the first day of the summer tourist season, due to record floods and rockslides generated by an extraordinary burst of torrential rain.

The Park will be closed to visitors, including those with reservations for lodging and camping, as officials assess the damage to roads, bridges, and other facilities, which is likely to last until Wednesday.

The closures come as Yellowstone approaches its 150th anniversary, and tourism-dependent companies in the area we're looking for a resurgence after two summers of COVID-19 travel restrictions.

For the first time since a series of deadly wildfires in 1988, all five park entrances were closed to approaching automobiles. The National Park Service announced that it was working to transfer visitors and personnel who had remained in various regions, particularly in Yellowstone's most severely affected northern section.

"The northern loop will very certainly remain closed for a long time," park superintendent Cam Sholly said in a statement.

According to the National Park Service, a mudslide to the north and washed-out road surfaces to the south cut off the "gateway" village of Gardiner, Montana, just beyond the park's northern boundary and home to many Yellowstone personnel.

Large sections of the winding North Entrance Road between Gardiner and park headquarters in Mammoth Hot Springs, Wyoming, were ripped away by swelling floodwaters along the Gardner River, according to aerial imagery supplied by the Park Service - washouts that will take months to fully restore.

According to officials, the park experienced widespread power outages, and preliminary investigations revealed that several routes encircling Yellowstone had been swept away or covered in boulders and mud, with a number of bridges destroyed.

Some roads in the park's southern reaches were on the verge of flooding as further rain was anticipated.

Days of torrential rain in the park, as well as continuous precipitation across much of the Intermountain West, contributed to the flooding and avalanche, which came after one of the region's wettest springs in many years. According to the park service, the rain and floods that swept through the park were "unprecedented," with the Yellowstone River exceeding its banks to new heights.

An unexpected surge in summer temperatures over the last three days has expedited the melting and flow of late-winter storm-related snow at the park's highest altitudes.

Just two weeks after the start of the summer tourist season in the United States, which accounts for the majority of Yellowstone's 4 million yearly visitors, heavy rains and high snowmelt flow occurred in the park.

Yellowstone National Park, the world's first national park, was established in 1872 and is now one of the most popular outdoor tourist destinations in the United States. It is known for its geysers, plentiful animals, and magnificent scenery, and it spans 2.2 million acres (890,308 hectares).

Floods in Yellowstone obliterate roads and bridges, stranding residents.

Flooding has washed away roads and bridges, cutting off all access to Yellowstone National Park just as the busy summer tourist season begins.

Due to the deluge, which was triggered by torrential rainfall and melting snowpack, all Yellowstone entrances were closed, and park officials hustled tourists out of the worst-affected sections.

There were no immediate reports of injuries, but scores of trapped campers in south-central Montana had to be rescued by raft. Authorities in Montana's Stillwater County stated they will be assessing a potential "loss of homes and structures."

Some of the biggest damage occurred in the northern area of Yellowstone National Park and the park's gateway communities in southern Montana, among other places. A landslide, a bridge washed down over a creek, and roadways badly undercut by churning floodwaters of the Gardner and Lamar rivers were all visible in images taken by the National Park Service in northern Yellowstone.

Gardiner, Montana, a village of around 900 inhabitants near the junction of the Yellowstone and Gardner rivers, just outside Yellowstone's busy North Entrance, was cut off from road access due to the flooding.

In the rushing Yellowstone River floodwaters just outside his door, tourist Parker Manning of Terra Haute, Indiana, got an up-close picture of the water rising and the river bank sloughing off at a cabin in Gardiner.

The Associated Press quoted Manning as saying, "We started seeing entire trees floating down the river, debris." "It was kind of insane to see one crazy single kayaker coming down through."

According to the National Weather Service, the Yellowstone River at Corwin Springs crested at 13.88 feet (4.2 meters) on Monday, breaking the previous record of 11.5 feet (3.5 meters) set in 1918.

Floodwaters flooded a street in Red Lodge, Montana, a hamlet of 2,100 people known for being the starting point for a picturesque, winding journey into Yellowstone's high country. Kristan Apodaca brushed away tears as she stood across the street near a washed-out bridge in Joliet, 25 miles (40 kilometers) northeast, according to The Billings Gazette.

The park where Apodaca's husband proposed flooded, as did the log home that belonged to her grandmother, who died in March.

"I am a sixth-generation American. "This is where we call home," she stated. "I literally drove across that bridge yesterday." My mother drove it before it was washed out at 3 a.m."

# CHAPTER THREE

## YELLOWSTONE FLOODING.

Yellowstone officials were evacuating the park's northern section, where routes could be unusable for an extended period of time, according to park Superintendent Cam Sholly.

However, the rest of the park was flooded as well, with park officials warning of even more flooding as well as potential problems with water supplies and wastewater systems in built areas.

In a statement, Sholly stated, "We won't know when the park will reopen until the floodwaters have retreated and we can inspect the damage around the area."

Officials say the park's gates will be sealed at least until Wednesday. The number of tourists removed from the site is unknown.

The rains arrived just as the tourist season for the summer began. One of Yellowstone's busiest months in June, which marks the start of an annual influx of nearly 3 million tourists that lasts into the fall.

Winter's remnants - in the shape of snow still melting and rushing off the slopes — made for a particularly awful moment for heavy rain.

Saturday, Sunday, and into Monday, Yellowstone received 2.5 inches (6 centimeters) of rain. According to the National Weather Service, the Beartooth Mountains northeast of Yellowstone received up to 4 inches (10 centimeters) of snow.

Cory Mottice, a meteorologist with the National Weather Service in Billings, Montana, stated, "It's a lot of rain "But the floods wouldn't have been anything like this if we hadn't had so much snow." "This is flooding as we've never seen before in our lives."

In the following days, the rain will likely stop, and lower temperatures will reduce snowmelt, according to Mottice.

Flooding on the Stillwater River stranded 68 people at a campground in south-central Montana. People were rescued by raft from the Woodbine Campground by Stillwater County Emergency Services organizations and employees from the Stillwater Mine on Monday. Due to flooding, certain routes in the vicinity have been closed, and inhabitants have been evacuated.

In a statement, the sheriff's office said, "When the waters recede, we will assess the devastation of homes and structures.

The deluge occurred while other parts of the United States were scorched by hot, dry weather. As a heat wave sweeps over states from the Gulf Coast to the Great Lakes and east to the Carolinas, more than 100 million Americans have been advised to stay indoors.

Crews from California to New Mexico battled wildfires in hot, dry, and windy conditions elsewhere in the West.

Climate change, according to scientists, is causing increasingly violent and frequent extreme events like storms, droughts, floods, and wildfires, yet specific weather occurrences are rarely directly connected to climate change without significant research.

This report was co-written by Associated Press writers Thomas Peipert in Denver and Mead Gruver in Fort Collins, Colorado.

The geology uncovered in Yellowstone, as well as the hidden characteristics revealed by scientific measurement, disclose the workings of the entire planet in an instantaneous and rarely witnessed manner.

But before you can read objective reports, you must first get beyond a lot of false sensationalism. This is how it's typically done: entice site browsers with a photo of Grand Prismatic Spring accompanied by either grossly false visuals or a sound depiction that warns when taken out of context.

Doesn't the right-hand image, along with a headline like the one above, imply there's a swelling chamber of lava, and melted rock, just beneath the park that's enlarging until it needs to explode above to wreak havoc? It's merely a matter of suggestion, with deceptive comments like "we don't know when it'll happen" implying we're on the verge of apocalypse. This is reckless "journalism," with non-evidence-based propaganda that draws unreasonable conclusions based solely on technical information.

What's really down there? On top of the increasingly hot, plastic mantle, there are kilometers of very hard rock atop a partially molten rock. Many fractures exist within the surface rock, from which water heated from beneath bubbles, steams, or spouts emerges. Even this following depiction is frightening, but it's meant to show that the entire body isn't filled with melted rock, but rather is only 10 volatiles, in a state where it may erupt if it were on the surface. After a new injection of hotter rock from below, it will take decades for a caldera-forming eruption to occur. Jesse though we've been told that the earth's skin is extremely thin at Yellowstone, it is nevertheless quite thick. The

6.4-kilometer-long blue bulbous bay on Yellowstone Lake's left shore is shown below. Depending on where you are, the only-partially molten magma beneath the park is five or more kilometers deep. Have you lately completed a 5-kilometer walk? In context, there's a lot of rock. When compared to most continental crust thicknesses of 30–60 km, it appears to be too thin.

My argument is that the magma isn't under pressure, isn't churning, and isn't about to burst. We're drifting over it and it's just sitting there. Before there is cause for alarm, a lot of changes must occur, leaving observable symptoms over a long period of time.

After reading Quoran's responses to Harry Turtledove's Supervolcano trilogy, I decided to read it myself. I understand why it isn't available at park bookshops. In a worst-case eruption scenario, he has a strong command of drama, which is crucial in illustrating the breakdown of practically all of the technical infrastructure we take for granted. However, his geographic and scientific errors, as well as social/governmental inaction in the aftermath of four years of significant crustal deformation (measured in feet per year in the book) and two minor volcanic eruptions, combined with hackneyed dialogue and inner monologue, make it nearly unreadable. However, he allowed for a two-year transition from normalcy to a mega eruption, whereas the 2005 telefilm "Supervolcano" only offered us a month.(At the very least, the protagonists of the teledrama were serious rather than kidding around.)

In the real world, a large eruption will be preceded by decades of bulging far greater than what is being measured, as well as a different type of earthquake than what is being recorded. That will serve as a warning.

According to a recent study, a mantle plume isn't required to heat Yellowstone's chambers; instead, micro subduction of a fragmented piece of plate is to blame.

Except for additional data suggesting indications of a deep mantle plume, with or without fractured Farallon fragments, this is all well and good. They're making assumptions based on tools that are far superior to those accessible before the fiction I've mentioned was created.

Below, whether feared or not, is the strength and promise of our planet. There's no reason why our people and resources can't be adjusted to best weather the unfathomable in the decades of warning signs that a changing magma chamber will provide. That is if it occurs in the next few years or millennia. (A modest event, on the other hand, is more likely than a planet-changing eruption.)

But, because scientists watching Yellowstone say there's still time, you may visit Yellowstone and Teton parks with confidence: it's not the end of the world. ™

We'll turn on the lava lamp for you. It is no city in Yellowstone National Park. Idaho, Montana, and Wyoming are all part of Yellowstone National

Park. With a total area of 2,219,791 acres (8,983.18 square kilometers), Yellowstone is more than three times the size of Luxembourg.

There's a surface response and a deeper one to this question.

For most people, Old Faithful, bears, and waterfalls are the first things that come to mind.

 the 26th of June, 2015, the 24th of June, 2019, and the 7th of September, 2014

But, as I see it, the Greater Yellowstone Ecosystem's core location, as the biggest intact biome in the contiguous United States, is of greater worth. Animals, plants, and other species are prospering rather than merely surviving as a result of a large amount of public space functioning as a buffer to the developed world.

Ironically, this is why, unless you're lucky and/or persistent, you're unlikely to encounter a bear, because they have so much terrain to explore that is out of sight of the road, and they are healthy and independent of humans.

Make no mistake: Yellowstone is primarily a geological park, with fauna as a secondary consideration. But because this region is unlike any other on the planet, this museum of how things used to be before technology incursion, and an exceptional attitude of preservation for all, particularly for its citizens and resources, it serves as an inspiration to all peoples.

It's famous for being dubbed "the super volcano destined to kill the world" by tabloid news outlets, but that's another story.

Yellowstone National Park is mostly in Wyoming's northern corner. The rest is in Montana's southwest and Idaho's southeast. Just to let you know, you can find this in just a second by running a google search.

How do you see Yellowstone National Park the best?

Yellowstone National Park is the most beautiful place I've ever been. It's as if you've entered another world. The Park is built on top of a volcano. The principal attractions are organized into four categories: hot springs, geysers, Yellowstone River, lower and upper falls, Yellowstone Lake, and wildlife in the Lamar Valley. The ideal way to see the park is to rent a compact automobile and stay in one of the park's many lodgings. You must select a lodging in accordance with your plans. As a result, I recommend entering by the north entrance near Mammoth Hot Spring. Visit Mammoth Hot Springs and wander around the walkways... (more)

Flooding in Yellowstone National Park damages a bridge and wipes out roadways.

On Monday, high water in the Gardiner River at Yellowstone National Park's North Entrance washed out a section of a road.

By way of AP, the National Park Service (NPS) is a government agency.

MONTANA — HELENA is a small town in the state of Montana. On Monday, major flooding at Yellowstone National Park took away at least one bridge, washed away roads, and triggered mudslides, causing officials to close the park's entrances and evacuate people.

The flooding was caused by "exceptional rains," according to park officials on Facebook.

Superintendent Cam Sholly said in a statement that "our first focus has been to evacuate the northern area of the park where we have several road and bridge failures, mudslides, and other difficulties."

Officials claimed the community of Gardiner, Mont., located north of the park, was cut off because the highways in and out of town were inaccessible. Some parts of the park are without power.

On Monday, flooding in Yellowstone National Park, Montana, swept down a bridge at Rescue Creek.

National Park Service via Associated Press

"We will begin to relocate people in the southern loop out of the park later today due to predictions of higher flood levels in portions of the park's southern loop, as well as problems with water and wastewater infrastructure," Sholly said.

Officials won't know when the park will reopen until the floodwaters have receded and they can survey the damage, he added.

"It is almost inevitable that the northern loop will be closed for an extended period of time," he predicted.

Officials claimed the park was witnessing unprecedented floods.

Scientists say they can't link a single weather event to climate change without more research, but climate change is causing more powerful and frequent extreme events including storms, droughts, floods, and wildfires.

Flooding was also caused by recent heavy rains and spring runoff in southern Montana, with water rushing down streets in Red Lodge on Monday. Carbon County has issued evacuation orders, according to the Office of Emergency Management.

Montana has been battling with water, while wildfires raged across the West in hot, dry, and windy conditions.

After previous floods, Yellowstone was closed; certain places were cut off.

The flooding occurred as a result of a torrential downpour paired with a quickly melting snowpack, and it occurred just as the summer tourist season was getting underway.

There were no initial reports of injuries despite the fact that countless homes and other structures were destroyed. Officials with Yellowstone said they were examining the damage caused by the storms, which washed away bridges, triggered mudslides, and required boat and helicopter evacuations.

It's unclear how many guests have been stuck or forced to leave the area, as well as how many residents who live outside the park have been rescued and evacuated.

The park's northern section and Yellowstone's gateway cities in southern Montana suffered the most damage. The National Park Service (NPS) is a federal agency Northern Yellowstone pictures showed a mudslide, washed-out bridges, and roadways undercut by the Gardner and Lamar rivers' swirling torrents.

Gardiner, Montana, a village of around 900 inhabitants near the junction of the Yellowstone and Gardner rivers, just outside Yellowstone's busy North Entrance, was cut off from road access due to the flooding. Floodwaters also cut off Cooke City, and residents in Livingston were ordered to evacuate.

FLOODS.

Park County officials, which includes those cities, announced on Facebook Monday evening that widespread flooding had made drinking water dangerous in numerous regions. Officials asked individuals who were in a safe area to stay put tonight as evacuations and rescues continued.

The Montana National Guard announced on Monday that two helicopters had been dispatched to southern Montana to assist with the evacuations.

Rain is not expected in the near future, according to Cory Mottice of the National Weather Service in Billings, Montana, and lower temperatures will reduce snowmelt in the coming days.

"This is flooding as we've never seen before in our lives," Mottice remarked.

Climate change, according to scientists, is causing increasingly violent and frequent extreme events like storms, droughts, floods, and wildfires, yet specific weather occurrences are rarely directly connected to climate change without significant research.

According to the National Weather Service, the Yellowstone River at Corwin Springs crested at 13.88 feet (4.2 meters) on Monday, breaking the previous record of 11.5 feet (3.5 meters) set in 1918.

Parker Manning got a close-up glimpse of the water rising and the river bank sloughing off in the rushing Yellowstone River flooding just outside his door at a cabin in Gardiner.

The Associated Press quoted Manning, who hails from Terra Haute, Indiana, as saying, "We started seeing full trees floating down the river, debris." "It was kind of weird to see one crazy single kayaker coming down through."

Manning stood by on Monday evening as surging waters undercut the opposite riverbank, causing a house to topple into the Yellowstone River and float away fairly undamaged.

Floodwaters flooded a street in Red Lodge, Montana, a hamlet of 2,100 people known for being the starting point for a picturesque, winding journey into Yellowstone's high country. Joliet, Illinois, is 25 miles (40 kilometers) to the northeast. The Billings Gazette reports that Kristan Apodaca wiped away tears as she stood across the street near a washed-out bridge.

The Park where Apodaca's husband proposed flooded, as did the log home that belonged to her grandmother, who died in March.

"I am a sixth-generation American. "This is where we call home," she stated. "I literally drove across that bridge yesterday." My mother drove it before it was washed out at 3 a.m."

Yellowstone officials evacuated the northern part of the park on Monday, citing the possibility that routes would be unusable for an extended period of time, according to park Superintendent Cam Sholly.

However, the rest of the park was flooded as well, with park officials warning of even more flooding as well as potential problems with water supplies and wastewater systems in built areas.

Saturday, Sunday, and into Monday, Yellowstone received 2.5 inches (6 centimeters) of rain. According to the National Weather Service, the Beartooth Mountains northeast of Yellowstone received up to 4 inches (10 centimeters) of snow.

Flooding on the Stillwater River stranded 68 people at a campground in south-central Montana. People were rescued by raft from the Woodbine Campground by Stillwater County Emergency Services organizations and employees from the Stillwater Mine on Monday. Due to flooding, certain routes in the vicinity have been closed, and inhabitants have been evacuated.

"In a statement, the sheriff's office said, "When the waters recede, we will assess the devastation of homes and structures."

The floods occurred as other parts of the United States were experiencing dry weather. More than 100 million Americans have been warned to stay indoors as a heat wave sweeps over states from the Gulf Coast to the Great Lakes and east to the Carolinas.

Associated Press reporters Thomas Peipert in Denver, Mead Gruver in Fort Collins, Colorado, and Lisa Baumann in Bellingham, Washington contributed to this story.

# CHAPTER FOUR.

## FLASH FLOOD

A FLOOD happens when water overflows or inundates normally dry ground.

Excessive rainfall, a dam or levee failure, or a quick release of water can cause flash floods in a matter of minutes or hours.

1. Relatively short in duration but high in intensity.

2. Has the ability to roll boulders, uproot trees, demolish structures and bridges, and scour new waterways.

3. Rainstorms that cause flash floods can also cause catastrophic mudslides.

4. Flash floods are responsible for the majority of flood-related deaths.

5. In steep and mountainous terrain, flash flooding is more likely than in level places.

I agree with Red Contreras' response.

• Flash floods are localized flooding that occurs suddenly.

Pay attention to the weather, and if you live in a flood-prone location, have a strategy in place.

Levies and storm sewers can help in some cases, but they all push the water to other locations, and as we spread out in our urban and suburban sprawl, there are fewer and fewer places to move the water without affecting someone.

Move homes out of flood-prone areas One challenge is that we filled in wetlands in many areas to build homes in the 1940s, 1950s, and 1960s. The wetlands served a dual purpose of giving the water somewhere to go and spreading out the flood so that if it did happen, it would be more diffuse and not as high. When wetlands are removed, the water has no place to go before it reaches people's homes.

Flooding is occurring more frequently and at higher crests as a result of climate change (regardless of who or what you believe is to blame), thus we must be aggressive in restoring wetlands as a buffer and encourage people to move away from flood plains.

Floods that occur in a "flash" are known as flash floods.

Consider how slowly water rises after a strong downpour. A river starts to surge, spilling over its banks and dispersing water away from the river. This is an example of a flood pattern.

Imagine a wall of water crashing down a gully or canyon. A sudden flood has occurred. They occur where the geography causes a bottleneck for moving water. The water from a large area uphill is channeled into a single location. A sudden flood has occurred.

- A series of cloudbursts have wreaked devastation in the western Himalayan area, wreaking havoc in Himachal Pradesh, the Union Territories of Jammu & Kashmir and Ladakh, and across the Pakistani border in Islamabad and Rawalpindi.

# Concerning Flash Floods

- Flash floods are short-duration, extremely localized phenomena with a very high peak, with fewer than six hours between the onset of rainfall and the peak flood.

- Flash floods differ from river floods in that they occur in tiny spatial scales and have a distinct character from river floods, making forecasting flash floods a different issue than standard flood forecasting systems.

- Because of its many hard surfaces, urban areas are more likely to encounter this form of "surface water" flooding.

When it rains, it doesn't soak into the earth as it does in the countryside.

- Consequences • It has an effect on the natural environment (including vegetation, agriculture, geomorphology, and pollution) and the human population (entrapments, injuries, fatalities).

- Issues • Flash floods are among the deadliest natural disasters in the world, claiming more than 5,000 lives each year and wreaking havoc on social, economic, and environmental systems. • They also have the highest mortality rate (defined as the number of deaths per several people affected) among the various types of flooding (e.g., riverine, coastal).

- Efforts in this direction

Guidance for Flash Floods It is a robust system designed by the India Meteorological Department (IMD) to provide the necessary products in real-time to support the development of flash flood warnings for the flash flood-prone South Asian countries of India, Nepal, Bhutan, Bangladesh, and Sri Lanka about 6-12 hours in advance at the watershed level.

- South Asian Flash Flood Guidance System (FFGS) • The South Asian FFGS was launched by the India Meteorological Department (IMD).

Its purpose is to assist disaster response teams.

- Assists governments in developing prompt evacuation preparations in advance of a flood.

Next Steps

- Each member nation must strengthen its weather observing networks so that more data is accessible when giving warnings; soil moisture data is also crucial, and the present network must be supplemented.

The Different Types of Floods

- It is the temporary flooding of huge areas caused by increased reservoir levels or rivers overflowing their banks as a result of torrential rainfall, high winds, cyclones, storm surge along the coast, tsunamis, melting snow, or dam bursts.

Floods that occur within six hours of the start of heavy rainfall are known as flash floods. Warnings for prompt evacuation may not always be possible in the event of flash floods. River floods are caused by heavy rains that fall across a vast catchment region. In contrast to flash floods, these floods usually build up slowly or annually and can last for days or weeks. Coastal Floods:

They are linked to cyclonic events such as hurricanes and tropical storms. • Catastrophic flooding is frequently exacerbated by wind-induced storm surges along the coast.

## What is the maximum height that a flash flood can reach?

The peak flow will be determined by two factors: the typical water channel's cross-sectional area (i.e., the area perpendicular to the flow direction) and the maximum discharge rate flowing through the channel.

In narrow ravines, mainly in the Alps, flash floods can reach heights of 20 meters. The influence of a river leaving the mountains on the open ground will be felt over a larger area. Perhaps the worst occurrence is the flooding of the Kedarnath valley when water levels surged more than 20 meters, or the flooding of old (and dry) Kosi river watercourses in North Bihar around a decade ago (following intense rainfall in Nepal and the release of impounded water).

# CHAPTER FIVE

## How quickly may flash floods happen?

*instantaneously*

There are numerous videos available on YouTube.

In the Texas hill country, several friends and I were camping on the bank of a trickling river. Rain was anticipated for the rest of the evening, although that isn't unusual for this time of year. The smoldering campfire was running out of fuel. I was in my tent when I heard movement.

The sound of a slow-moving freight train could be heard. Trains were widely used.

Suddenly, a wall of steam from the fire engulfed my tent, filling it with unbearable heat. I was suffocating, and as I tried to get out, a football-sized rock slapped me in the shin, as if I had walked into a trailer hitch. Scar tissue image on demand

As we saw all of our shit go away, we all congregated on higher land, 10 feet up.

3 minutes total elapsed time

To keep safe from the torrent, we had to back up almost 100 yards from our original campground during the following hour.

We were lucky in that there was plenty of space for us to move around.

# What caused the flash flood to occur?

"How do flash floods happen?" I assume you want to know.

Heavy rain, dam failure, quick snowmelt, and other factors can cause flash floods. A flash flood is more likely to occur within a few hours of the onset of a severe downpour. It's dubbed a 'flash' flood because it happens suddenly. It can also happen when rainfall surpasses the capacity of the ground to absorb the water in the area. Because houses, roads, and driveways make a concrete jungle, increasing water runoff by lowering the amount of rain that the land can absorb, highly populated areas are in danger of flash floods.

When the depths of water in a river or rivulet swiftly rise above the banks, it is called a flash flood. Flash floods are more likely to cause property damage in locations built along or near a stream or river. Furthermore, continuous heavy rain on slopes can destabilize the soil and generate mudslides or soil landslides, which can cause property damage. When rainstorms move swiftly, flash floods are infrequent because the rainwater is scattered across a large area. Flash floods can occur when slow-moving or persistent precipitation falls over a vast area.

Flash floods can also be caused by dam failures. When a dam is breached because of a significant amount of water, a devastating e wall of water is created, destroying everything downstream.

Flash floods can be found in the Sahara Desert's west, center, and east regions, which may surprise you.

**What was the most devastating flash flood you've ever witnessed? What are the most common locations for flash floods?**

(1) The one that occurred in Eastern Oregon and Washington, resulted in the 'Channeled Scablands' by stripping the land down to bedrock after a catastrophic ice dam collapse released the waters of the giant glacial Missoula Lake down the Columbia River drainage, or (2) the one that occurred in Eastern Oregon and Washington, which resulted in the 'Channeled Scablands' by stripping the land down to bedrock by releasing the waters of the giant glacial Missoula Lake down the This has happened numerous times, the most recent being nearly 14,000 years ago.

(2) The 7100-year-old inundation of the Black Sea by the Mediterranean Sea.

(3) The contentious Zanclean flood, occurred around 5 million years ago when the straits of

Gibraltar opened and flooded the Mediterranean region with water from the Atlantic Ocean.

What should you do if you find yourself in the middle of a flash flood?

As quickly as possible, you must ascend to higher land. By whatever means necessary, stay out of the water.

Now, your odds of getting harmed or dying vary drastically based on the type of terrain you're on. If you're in a gully when the flood comes, and the terrain is generally flat or undulating, you'll have a good chance of getting to higher land.

You may have no choice but to climb if you're in a narrow canyon with steep or sheer walls. The deadliest canyons are those that are extremely narrow-slot canyons. You might not make it if you get trapped in a flood in one of these. In terms of planning your trip, this is where risk management comes into play.

Find out how big the watersheds that feed your canyon are. You might need to know everything you can about weather conditions and forecasts a hundred miles or more away. Floodwaters may take hours to reach you if severe rain hits. It's also possible that the sky you can see is clear and sunny.

I no longer walk into canyons with small watersheds since I can see all of the lands that will provide water to my chosen canyon.

There are no promises, however. Every day, over 1,800 thunderstorms occur around the world. That doesn't seem like a lot for our great, magnificent globe, at least to me. However, one in your chosen watershed is all it takes to ruin a beautiful day in the canyon.

This video shows a flash flood in Arizona's Antelope Canyon. The canyon is about 120 feet deep here, and floodwater is spilling over the crest, as you can see.

## Is it possible to foresee flash floods?

Because flash floods are not always driven by meteorological processes, they pose unique forecasting and detection issues. When favorable meteorological and hydrologic conditions combine, flash floods occur. Although heavy rain is required, depending on the hydrologic characteristics of the watershed where it is raining, a given amount and duration of rain may or may not result in a flash flood. • knowing how much water runs off (and where it goes) • knowing how strong the stream is flowing • knowing how wide an area is getting rain • knowing how hard and fast it is raining • knowing how long it has been raining in a particular drainage basin • knowing where the storm is located and how fast or slow it is moving • knowing how porous the soil is and how much water it already holds • knowing how much vegetation covers the soil • knowing how much surface is paved

Hydrologists use gauges to measure the water levels in streams, rivers, and lakes. Hydrologists research the effects of water on the earth's surface and in the atmosphere. Snow gauges are also used to determine the amount of water in the snow. They take into account the amount of rain that has fallen recently (since soil moisture impacts how much rain soaks in and how much runs off), as well as the amount of rain that meteorologists forecast to fall in the future. The information is delivered to a river forecast center, where computers analyze it to forecast river and stream levels in the area. When local forecasters receive the information, they compare it to charts for their area and, if required, issue a flood warning.

## What causes flash flooding in the desert?

Every summer, intense cloudbursts that can drop one or two inches of rain over many square miles are a typical occurrence in the Southwest. From Mexico, large swaths of moist air move north. The monsoons are how people in New Mexico, Arizona, Nevada, and Southern California refer to them. The hot desert floor warms the air, pushing the wet air above it into the stratosphere, where it is a hundred degrees below zero. Rain condenses from the moist air and falls in a downpour. The desert lacks a dense network of plant life to absorb the rain, so it flows off into gullies.

We saw them every summer when I lived in Green Valley, Nevada, and for an hour or two, a wide gully behind the cul-de-sac that was twelve feet deep and sixty feet broad would partly fill with fast-moving water. Our street's storm drain manhole covers would pop, causing geysers to erupt.

This water rushes into gullies, filling them quite quickly. Every few years, Caesar's Palace would pave its parking lot into the gully adjacent to it, causing a hundred cars to be flooded and many to be swept away.

In a matter of hours, these flash floods have the power to carve vast valleys or shift their path. A hundred feet behind their home, friends in North Las Vegas had a twenty-foot gully. After the gully filled with rushing water, it was cut straight up to their house.

Thunderstorms and microbursts can be seen in the valley's higher levels during the monsoon season in Las Vegas. They squirt a large amount of water onto a small area in a short period of time. This is how things appear to be.

Take a stand! if you're stuck in a flood!! A Flood Warning is issued when a dangerous weather event is impending or has already occurred. A Flood Warning is issued when flooding is impending or has already occurred

What are some warning indicators that a tornado is approaching in the absence of an official emergency alert?

It's tough to breathe when the weather is oppressively hot and humid, and the heat appears to last forever.

Then it begins to rain, with unexpected gusts of wind, and the temperature begins to drop.

As they gaze out the window at the rain, someone remarks, "Perhaps this will wipe some of the humidity out of the air..."

This is something I learned in Georgia, when my neighbors began listening to weather forecasts at the same time I was relieved by the change in weather. It happened once outside of DC... a hot, still day with swamp-thick air that wouldn't move, then powerful gusts of wind and rain... A tornado warning was issued for my county a half-hour later.

The best technique I know of is to listen to local radio stations, particularly public radio stations. Keep an eye on the weather channel as well.

Simply because a local stream is rising does not mean it will continue to rise or will reach a dangerous level. If you reside in a flood plain, it's best to be vigilant all of the time.

In some circumstances, contacting the local authorities can be advantageous.

## What does an Areal Flood Warning entail?

An Areal Flood Warning is issued when flooding comes gradually, usually as a result of prolonged and continuous moderate to severe rainfall. As a result, low-lying, flood-prone areas, as well as small creeks. Even while this type of flooding takes longer than flash floods to form, it can still be dangerous to people and property.

## What precautions should you take if a storm strikes without warning?

Storms don't just show up out of nowhere. Threat forecasts are released by each US office days in advance.

As a result, the following actions need to be taken:

1. Make a PLAN that takes the worst-case situation into account. Have you ever been affected by earthquakes, wildfires, tornadoes, or tsunamis? What if a solar storm occurs? It's possible that it will happen again if it has happened previously.

2. Put together a kit to go with the plan. Are you in need of cash, food, replacement medicine, or eyeglasses? Is your car's gas tank full enough to get you away? Which path will you choose? Where are you going to spend the night?

3. Carry out the strategy. Did it function when there was no stress? If it didn't, it wouldn't work on a bad day. Rewrite the parts of the plan that

didn't go as planned. Ensure that your emergency supplies are up to date.

4. Assist your friends and neighbors if they need it. You are only capable right now.

Flood-prevention systems

## What Is the Importance of Monitoring?

While some areas are more susceptible to flooding than others, flood warning systems installed near any major waterway or body of water provide vital information that can save property and lives. The most efficient flood warning systems, in fact, go beyond establishing gages and telemetry equipment to employing qualified individuals and well-designed protocols to provide the earliest possible notice regarding whether a flood is expected, when it will occur, and how severe it will be. This resource can be used by individuals, communities, and organizations interested in building and operating flood warning systems.

# CHAPTER SIX

## Flood Warning.

The National Oceanic and Atmospheric Administration's US Geological Survey and the National Weather Service work to keep flood warning systems operating across the United States. The United States Geological Survey (USGS) is the principal source of surface and groundwater data, and it operates more than 85% of the country's stream gaging stations. The NWS uses these and other data sources to generate river forecasts and flood warnings.

The NWS issues flood warnings on a county-by-county basis, as well as for specific rivers and streams. There are several basic sorts of these alerts:

When there is a chance of flooding or flooding is likely within the next 12 to 48 hours, flood watches are issued.

Flood warnings are issued when widespread flooding is expected over a large area, or if flooding is imminent or already happening.

Flash flood watches and warnings are issued in the same way as conventional flood watches and warnings, but they indicate the likelihood of exceptionally rapid flooding, which is typically triggered by heavy rain or dam failure.

When big streams are expected to flood, flood warnings are issued, but people and property are not in risk. They could also be used as a follow-up to previous alerts and warnings.

These alerts are broadcast in Specific Area Message Encoding by the Emergency Alert System and the NOAA Weather Radio network in the United States.

In towns that do not have a flood warning program but are interested in developing one, the NWS can provide extra guidance and technical support, as well as outreach and education to relevant parties and community leadership. The advantages of a flood warning system, which include the protection of lives and property, far outweigh any potential complications or inconveniences.

At every stage of the process, Fondriest Environmental can assist you with the installation and maintenance of gages, sensors, and other equipment.

## *A Response in Real-Time*

For an effective flood warning system, data on local rainfall, stream level, and streamflow must be collected on a regular basis. This can be done through routine monitoring, which includes visits by operating personnel to stream gage and precipitation measuring sites, but a real-time monitoring system with telemetry can make data collection easier — and, in many cases, more cost-effective — while allowing for the fastest possible response to a flood event. According to the NWS, even in areas where the NWS offers flood warning coverage, a real-time, community-oriented flood warning system can reduce the dangers of flooding.

Floods are predicted by the National Weather Service (NWS) using complex mathematical models that estimate how rivers and streams across the United States will react to varying levels of rainfall and snowmelt. If you wish to build a responsive flood warning system without advanced forecasting skills, you can probably get by with a system based on Automated Local Evaluation in Real-Time, or ALERT gages.

### *Flood Warning System with Automated Alerts*

Data collection via gaging, data processing, and the necessary hardware and software, as well as flood warning information transmission, are all crucial parts of establishing a flood warning system. While automated flood warning systems are often affordable to install, the number of gage site locations is the most critical element in determining the cost. The type of communications and telemetry capabilities available at each location will also affect costs.

ALERT Gages

There are several different types of automated stream gages that may broadcast stream-level data via telemetry, but those that follow the NWS ALERT protocol are the most common and the focus of this tutorial. Many other gages used to measure precipitation and water level work on the same principles like this one, and some of the information in this article may be applicable to other systems.

ALERT systems benefit from a defined set of communication protocols, which means that while a variety of firms design and manufacture ALERT hardware and software, the vast majority of those products are interchangeable.

The basic roles of ALERT gages are sensing and communication. ALERT gauge sensors track changes in a specific parameter, most often precipitation volume and/or water level. In more modern gages, temperature and wind speed sensors may be included. Some ALERT gages can also provide information about the health of the unit or site-specific data.

A precipitation-measuring ALERT system, for example, will be configured to detect a certain "event," such as 1 millimeter of rainwater entering the gage's tipping bucket through the top of its funnel. When the bucket tips, any water inside is poured out, activating a switch that sends ALERT data and resetting the bucket. After detecting a certain incident, any additional sensors on the gauge will also trigger the ALERT data transmitter. ALERT gages will send a "no rain" signal on days when there is no rain to show that the device is still operational.

### *Processing of Information*

Depending on the user's needs and preferences, different applications will be used to gather and process data from ALERT gages. Many ALERT gage makers have their own proprietary software that may be used to view data remotely in graphical or text forms. Several people can access the data at the same time, and multiple gages can be monitored at the same time using the most useful ALERT processing software.

## *Publication of Information*

Automated flood warning systems can connect with a host computer or network by radio, cellular, or satellite telemetry, however, ALERT systems only use radio frequencies. As a result, ALERT systems are susceptible to the same problems as any other radio transmission device, such as interference from electrical noise and environmental conditions. Interference can also arise if multiple ALERT systems communicate at the same time in close proximity. Although satellite and cellular telemetry tend to avoid these issues, site selection must still be considered in order to enhance transmission quality. A power supply will be required for all types of automated flood warning systems. While telemetry devices in close proximity to developed settlements may be powered by a commercial power system, those in remote locations are often fueled by a combination of battery and solar power.

## *Measurements of Streamflow*

While streams and rivers share many characteristics and parameters with lakes, ponds, and basins, they have one feature that distinguishes them from other freshwater bodies: movement. Streamflow is a critical element that influences the hydrology and water quality of a river. Although these other components of a river's health may be just as important — or just as relevant to your project - they may be shared with other types of water bodies and, in many cases, will be explored in subsequent chapters. As a result, the focus of this chapter will be on determining streamflow through stage-discharge measurement.

More information can be found in the section on Streamflow Measurements.

## *Flood Warning System in Practice*

As previously mentioned, there are several ways to configure an automatic flood warning system, but the requirements of one system may differ significantly from those of another. Depending on the nature of your application and the extent of the targeted coverage area, the number of gage sites, their positions, and the equipment and sensors utilized at each will vary. The number of gages required for a community-wide warning system will be determined by the placement of local water bodies in relation to property and infrastructure. One gauge may be sufficient if only a tiny area of your community is exposed to a protruding stretch of river, for example.

Installing a station on a riverbank or a standing structure, such as a pier or bridge support, will likely produce the greatest results in a single-gage system. Gages can also be installed inside stilling wells or standpipes, making it easy to incorporate other instruments like multi-parameter sondes with many sensors, data loggers, and telemetry systems. While radio transmission is the conventional telemetry option for ALERT-based systems, depending on the size and location of your application, satellite and cellular solutions may be more helpful. Almost all telemetry options will provide continuous real-time data to any computer or mobile device, guaranteeing that your system functions smoothly and that any control or emergency actions may be taken quickly if parameter restrictions are exceeded.

### *Rain Gage in a Tipping Bucket*

Rain gauges are made up of a funnel and a tiny container attached to a tipping lever that gathers a specific amount of precipitation before tilting the container, discharging any collected water, and transmitting an electrical signal to a data transmitter.

The Integrated Data Logging System (IDLS) is a system that allows you to

A real-time monitoring station that houses the data logger, telemetry module, and power/charging supply is known as an integrated data logging system. Because connecting AC power to the monitoring position is typically prohibitively expensive, integrated solar panels are employed to continuously charge the 12VDC battery for autonomous operation.

## *Hardware for Mounting*

Depending on the location and recommended monitoring plan, data loggers can be fixed directly to the pier/abutment or to a pole on the bridge.

## *Water Level Radar Sensor*

Radar-based water level sensors offer a non-contact alternative to traditional level gauging technologies such as submersible pressure transducers, enabling monitoring in difficult-to-reach areas.

Telemetry allows users to obtain data in real-time. ALERT uses radio frequencies to provide wireless communications, although cellular and satellite-based solutions are also available.

## *Data that is constantly updated*

A cloud-based data center provides instant access to project data 24 hours a day, 7 days a week. To spot trends, monitoring data can be examined in real-time or as a graph. When certain parameters surpass predefined limitations, real-time automated notifications can be issued by SMS or email.

### *Location Surveillance*

As previously stated, the best location for a flood warning gage will be determined by the site considerations of the canal in question. Selecting the location, determining substrate stability and water level fluctuation, and designing a housing solution that would adequately protect the gage from acts of nature or vandalism all require careful preparation. When choosing a place, take into account the physical limits of the area, the time it takes to get there, legal and physical access to the property, and safety concerns.

The radar sensor should be firmly mounted on a bridge or above the structure at the monitoring position. To avoid misleading reflections, there should be a clear passage between the sensor and the water.

If you're utilizing telemetry, check to see if there's enough cellular coverage at the location to receive a signal. Otherwise, satellite telemetry will almost certainly be required.

Avoid submerged objects that alter or influence the water level, such as rocks or bridge piers. When the water is at its lowest expected level, look for such impediments.

o Horizontal structural surfaces such as beams, brackets, and side wall joints should be avoided while mounting since they tend to reflect a high misleading signal. Avoid man-made control/flow structures upstream or downstream of the site that could cause erroneous flow profiles, as this will complicate rating development.

## *Considerations for Installation*

Ensure that data logging and telemetry equipment is situated high enough above the high-water mark to avoid flooding or other damage from high water.

o The radar sensor should be placed exactly above the water's surface, with the radar beam perpendicular to the water's surface.

· Wind and vibration should not cause the radar sensors to move vertically, thus they should be securely fixed. Any movement can cause mistakes in measurements and vertical alignment.

Make sure the radar sensor is mounted high enough to avoid getting flooded during floods or high water.

To protect against animals, all potentially exposed sensor cables near the ground should be run through a conduit to the data logger box.

## *Management of data*

Collecting and analyzing data is an important aspect of every monitoring activity. While taking measurements on site is conceivable, the ability to log, send, and evaluate monitoring data in real-time is significantly more efficient.

In the Monitoring Equipment section, you may learn more.

## Assurance of Quality

Cleaning and calibrating instruments at regular intervals is recommended to preserve the accuracy and keep equipment running within specifications. It's also a good idea to double-check sensor accuracy using a different instrument. Projects may even necessitate the adoption of a Quality Assurance Plan (QAP), which outlines the maintenance, calibration, and QA/QC requirements in detail.

## Up keep of the system

Radar sensors require almost no upkeep. There are no calibration intervals or consumable parts to worry about. However, it is still necessary to visit the site on a regular basis to check for potential issues.

Examine the sensor for dirt, spider webs, insect nests, and other obstructions that could affect the measurement. If any obstructions are found, clean the sensor gently with a non-abrasive cleaner and a soft sponge.

It's also crucial to inspect the measuring beam for any obstacles. Flotsam, as well as branches from trees and plants growing in the water, can be found here. Remove any obstacles in the way of the beam.

## *Verification of Results*

It's also necessary to ensure that the sensor is delivering reliable data on a regular basis, in addition to visually inspecting the sensor and measurement beam. A portable distance sensor or a nearby staff gauge can be used to do this.

## *Equipment that is recommended*

There are numerous pre-made ALERT systems available for purchase when it comes to building an automatic flood management warning system, but tailoring a system to your individual demands may deliver the greatest results. For their quality, reliability, and value, Fondriest Environmental has chosen these items as the best in their category. They work together to give an advanced and robust real-time monitoring system for any prone-to-flooding river or stream. The OTT RLS Radar Water Level Sensor uses radar pulse technology to measure depth in situations where contact-based depth sensors are not suitable. The HSA TB3 Tipping Bucket Rain Gauge provides accurate rainfall data with a margin of error of less than 3% for rainfall intensities ranging from 0 to 500 millimeters per hour. The NexSens 3100-MAST Wireless Telemetry Device combines a mast-mounted data logging system with cellular modem telemetry and solar charging to keep your data current, removing the need to visit a gage site on a regular basis. In addition, the WQData LIVE web datacenter provides remote access to collected data from any computer or mobile device 24 hours a day, seven days a week, with fast alert notifications and trend tracking.

# CONCLUSION

## YELLOWSTONE RIVER

The Yellowstone River, which flows 671 miles (1080 kilometers) from its source southeast of Yellowstone to the Missouri River and subsequently to the Atlantic Ocean, is the last major undammed river in the lower 48 states. It begins atop Yount Peak in the Absaroka Mountain Range. The river enters the park and flows into Yellowstone Lake via the Thorofare region. It emerges from the lake near Fishing Bridge and flows north into Hayden Valley, passing via LeHardys Rapids.

It crashes over the Grand Canyon's Upper and Lower Falls after this tranquil stretch. It then flows northwest until it reaches Tower Junction, where it meets its largest tributary, the Lamar River. It leaves the park near Gardiner, Montana, after passing through the Black Canyon. Montana's Yellowstone River travels north and east before entering the Missouri River near the state's eastern border. The Missouri River eventually merges with the Mississippi, which empties into the Gulf of Mexico and the Atlantic Ocean.

Many of the spawning streams in the Lake Village, Fishing Bridge, and Bridge Bay area's provide essential food sources for grizzly bears in the spring, in addition to the Yellowstone River. As a result, these rivers and streams represent a primary resource in the district from an ecological standpoint. Three miles north of Fishing Bridge, the LeHardys Rapids are a cascade on the Yellowstone River. According to geomorphology, this is the exact location where the lake terminates and the river continues northward. Many cutthroat trout can be found here in the spring, resting in the shallow pools before bursting into bursts of energy to sprint up the rapids on their route to spawn under Fishing Bridge.

Paul LeHardy, a citizen topographer with the Jones Expedition in 1873, was honored with the rapids' name. Jones and a companion set off on a raft to examine the river, intending to meet up with the remainder of their party at the Lower Falls. The raft capsized when it hit the rapids, and many of the supplies, including firearms, bedding, and food, were lost. LeHardy and his companion saved what they could and walked the rest of the way to the falls.

About In the lower 48 states, the Yellowstone River is the last large free-flowing river. The Yellowstone River begins high in Yellowstone National Park's interior, at Yellowstone Lake in Wyoming, and flows 676 miles to its confluence with the Missouri River, unrestricted by a single dam. The Yellowstone River eventually enters the Missouri River just past the North Dakota border. The Yellowstone River continues as it has for millennia, undammed and wild, from its headwaters in Lake Yellowstone downstream 670 miles to the Missouri River in North Dakota. It offers fishers and river runners some of Montana's best water: clear, cool, rapid, and difficult. Its blue-ribbon rivers are famous for their quantity of large trout among anglers all over the world. The Big River passes through various variations as it travels across Montana, from steep-walled gorges where rocks churn its green waters to the eastern section of the state, where it broadens and takes a languid meander over fertile farmland. Montana agates are frequently found in the gravel along the river from Custer to Sidney (inquire locally or through the Custer Country tourism region for guide services or rock hounding tips). Fishing and floating are two more popular Yellowstone activities. Badland's landscape may be found in abundance in Terry or Makoshika State Parks (both in Glendive), where you can travel to see incredible rock formations.

www.ingramcontent.com/pod-product-compliance
Lightning Source LLC
Chambersburg PA
CBHW051539240526
45465CB00028B/1556